Worksheets

Exploring Drafting

11th Edition

John R. Walker
Bernard D. Mathis

Publisher
The Goodheart-Willcox Company, Inc.
Tinley Park, Illinois
www.g-w.com

Introduction

The worksheets in this student supplement enrich and reinforce the material presented in *Exploring Drafting*. They are intended to help you develop problem-solving ability. Problems selected from the text are presented in order of increasing difficulty. Using these worksheets helps to eliminate repetitive drawing, and enables you to get directly into problem-solving situations.

Drawing numbers shown to the right of the title blocks key the drawing to chapters in *Exploring Drafting*. For example, information that will aid in solving Problem 6-1 can be found in Chapter 6. Blank worksheets may be used for special assignments.

John R. Walker
Bernard D. Mathis

Contents

		Textbook Page	Worksheets Page
1	Why Study Drafting?	17	
2	Careers in Drafting	27	
3	Sketching	41	5
4	Drafting Equipment	67	
5	Drafting Techniques	91	23
6	Basic Geometric Construction	121	37
7	Computer-Aided Drafting and Design	151	
8	Lettering	187	55
9	Multiview Drawings	203	63
10	Dimensioning	235	107
11	Sectional Views	273	119
12	Auxiliary Views	297	139
13	Pictorials	311	167
14	Pattern Development	349	189
15	Working Drawings	371	205
16	Making Prints	387	
17	Design	399	211
18	Models, Mockups, and Prototypes	411	
19	Maps	423	215
20	Graphs and Charts	437	221
21	Welding Drafting	451	229
22	Fasteners	465	239
23	Electrical and Electronics Drafting	481	243
24	Architectural Drafting	493	249
25	Manufacturing Processes	527	

Sketch angular lines.

Sketch a circle.

Sketch horizontal and vertical lines.

Sketch a square.

Problem
3-1a

Date

Name

Problem

Date

Name

Sketch vertical lines.

Sketch horizontal lines.

Sketch angular lines.

Sketch angular lines.

Problem
3-1b

Date

Sketching Lines

Name

Problem

Date

Name

Sketch a 30°-60° angle.

Sketch a 45° angle.

Sketch a hexagon.

Sketch an octagon.

Sketch a square.

Sketch a circle.

Sketching Shapes and Angles

Name

Date

Problem

Date

Name

Construction Line

Border Line

Phantom Line

Hidden Line

Centerline

Cutting-Plane Line

Object Line

Dimension Line

Section Line

Sketch each type of line.

Alphabet of Lines

Name

Date

Problem

3-2

Problem

Date

Name

Sketch a problem assigned by your instructor.

Sketching Problem

Name

Date

Problem

Date

Name

Sketch a problem assigned by your instructor.

Sketching Problem

Name

Date

Problem

Date

Name

Sketch a problem assigned by your instructor.

Sketching Problem

Name

Date

Problem

Date

Name

Sketch a problem assigned by your instructor.

Name

Sketching Problem

Date

Problem

Date

Name

Sketch a problem assigned by your instructor.

Sketching Problem

Name

Date

Problem
3-7

Problem

Date

Name

Alphabet of Lines

Name

Date

Construction Lines

Border Lines

Phantom Lines

Hidden Lines

Centerlines

Cutting-Plane Lines

Object Lines

Dimension Lines

Section Lines

Problem

Date

Name

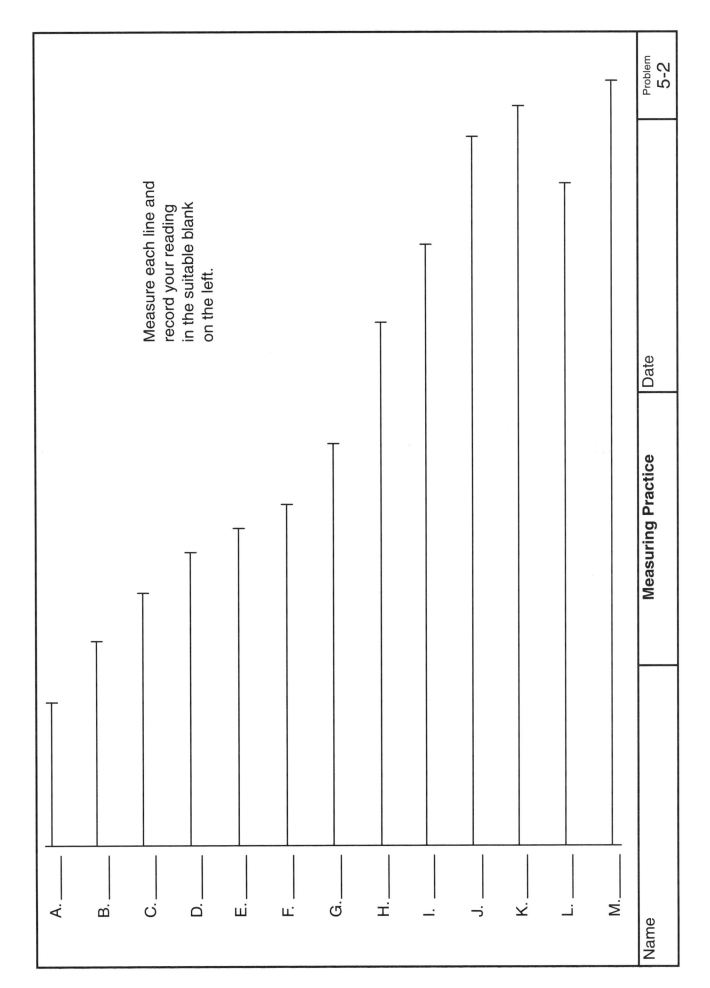

Measure each line and record your reading in the suitable blank on the left.

A. ____
B. ____
C. ____
D. ____
E. ____
F. ____
G. ____
H. ____
I. ____
J. ____
K. ____
L. ____
M. ____

| Name | | Measuring Practice | Date | Problem 5-2 |

Problem

Date

Name

Draw horizontal lines. Space lines at 3/8".

Draw vertical lines. Space lines at 1/2".

Draw 45° lines to the right. Space lines at 1/2".

Draw 45° lines to the left. Space lines at 1/2".

Instrument Practice

Name		Date

Problem

Date

Name

Draw 60° lines to the left.

Draw 30° lines to the left.

Draw 60° lines to the right.

Draw 30° lines to the right.

Problem
5-4

Instrument Practice

Date

Name

Problem

Date

Name

Draw 75° lines to the left.

Draw 15° lines to the left.

Draw 75° lines to the right.

Draw 15° lines to the right.

Instrument Practice

Name

Date

Problem
5-5

Problem

Date

Name

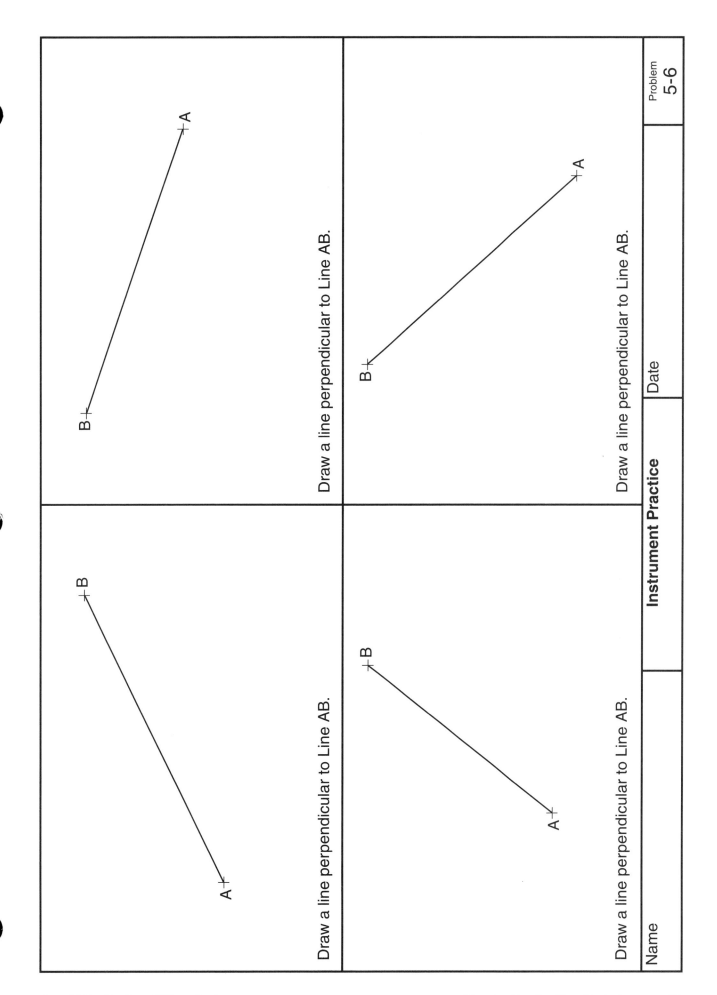

Draw a line perpendicular to Line AB.

Draw a line perpendicular to Line AB.

Draw a line perpendicular to Line AB.

Draw a line perpendicular to Line AB.

Instrument Practice

Name

Date

Problem
5-6

Problem

Date

Name

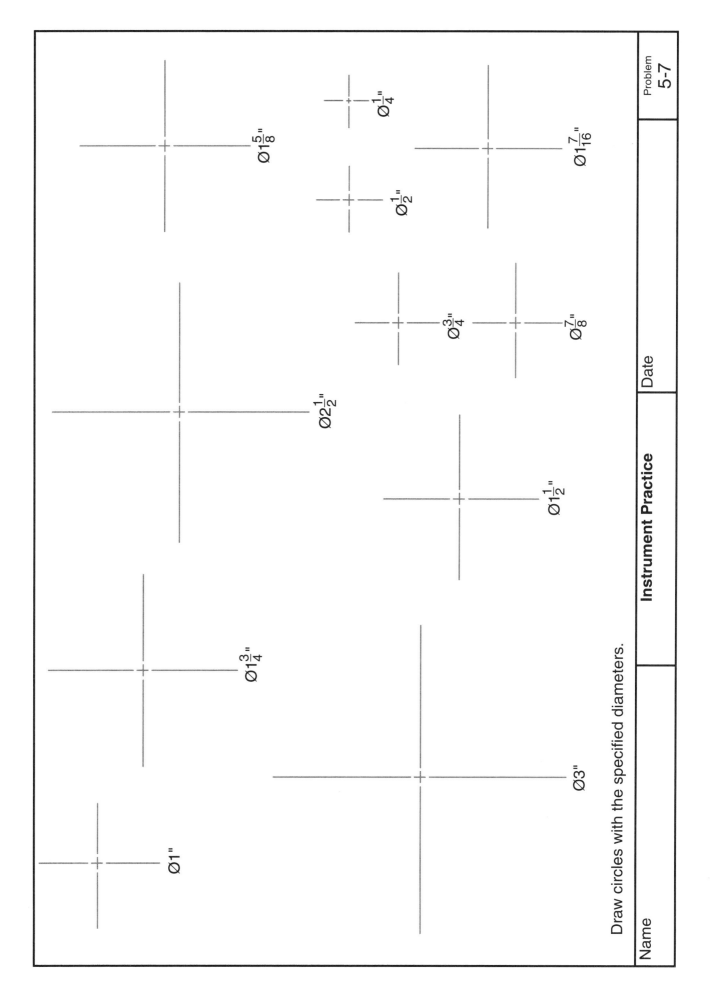

Draw circles with the specified diameters.

Name

Instrument Practice

Date

Problem
5-7

Problem

Date

Name

Bisect angle ABC.

A

B

C

Bisect arc AB.

A

B

Bisect line AB.

A

B

Construct an equilateral triangle.

A

B

Construct a triangle.

A

B

C

C

Transfer angle ABC.

A

B

C

A'

B'

Geometric Problems

Name

Date

Problem
6-1

Problem

Date

Name

Construct a pentagon.

Inscribe a hexagon.

Construct a square given a 2" side.

Construct a hexagon using the 30°-60° triangle.

Construct a square given a 2½" diagonal.

Construct a hexagon using the compass.

Geometric Problems

Name

Date

Problem

Date

Name

Divide Line AB into nine equal parts.

Draw an arc tangent to Lines AB and BC.

Construct an octagon.

Draw an arc tangent to Lines AB and BC.

Construct an octagon.

Draw an arc tangent to Lines AB and BC.

Geometric Problems

Name

Date

Problem
6-3

Problem

Date

Name

Circle = Ø2"
Arc = R$\frac{1}{2}$

Draw an arc tangent to a 2" diameter circle and the straight line.

Circles = Ø2"
Arcs = R1

Draw arcs tangent to two 2" diameter circles.

Name	**Geometric Problems**	Date	Problem 6-4

Chapter 6 Basic Geometric Construction **43**

Name		Date	Problem

Draw an ellipse using the concentric circle method.

Major axis = $5\frac{1}{2}"$
Minor axis = $4"$

Ellipse Construction

| Name | | Date |

Problem

Date

Name

Draw an ellipse using the parallelogram method.

Problem
6-6

Ellipse Construction

Date

Name

Problem

Date

Name

Draw an ellipse using the four-center approximate method.

Problem 6-7

Name

Date

Ellipse Construction

Problem

Date

Name

Draw a United States aircraft insignia with a 4" diameter star.

Design Problem

Name

Date

Problem

Date

Name

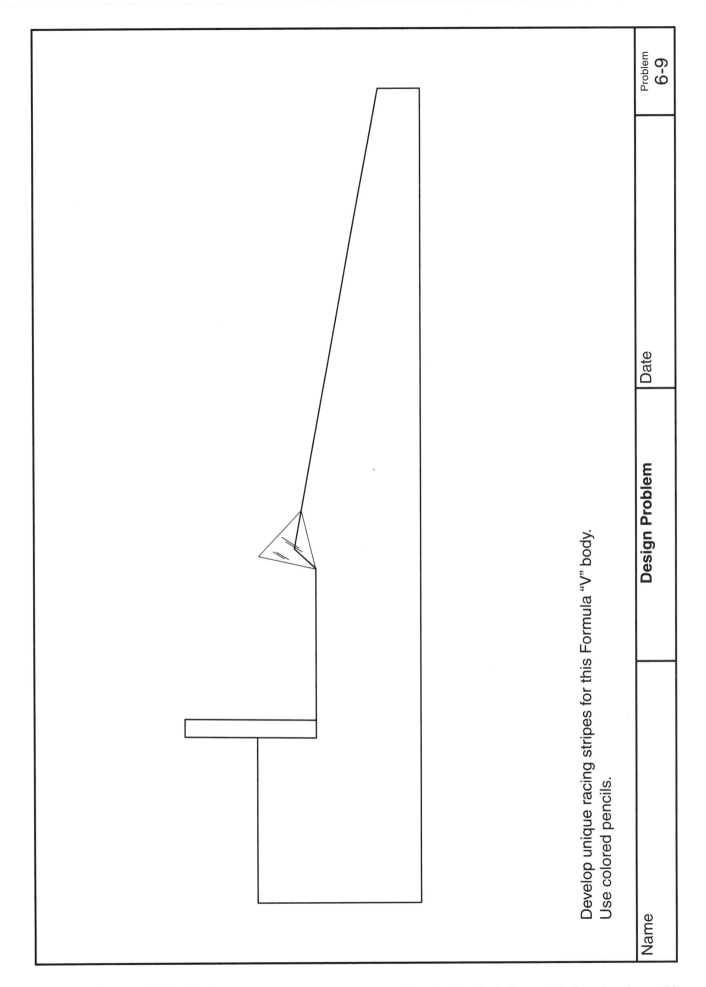

Develop unique racing stripes for this Formula "V" body. Use colored pencils.

Date

Design Problem

Name

Problem

Date

Name

A GOOD DRAFTER LETTERS
NEATLY AND RAPIDLY.

YOUR NAME
YOUR SCHOOL

Date

Lettering Practice

Name

Problem

Date

Name

THE QUICK RED FOX JUMPED

OVER THE LAZY BROWN DOG.

1 2 3 4 5 6 7 8 9 0 $\frac{1}{2}$ $\frac{1}{16}$ $\frac{3}{8}$ $\frac{5}{32}$

Name

Lettering Practice

Date

Problem

Date

Name

"THE BEST THING ABOUT THE
FUTURE IS THAT IT COMES ONE
DAY AT A TIME."

MAKE THE MOST OF YOURSELF
FOR THAT IS ALL THERE IS OF
YOU.

Date

Lettering Practice

Name

Problem

Date

Name

PACK EACH BOX WITH SEVEN

DOZEN GIANT JUGS.

1 2 3 4 5 6 7 8 9 0 $\frac{1}{2}$ $\frac{3}{4}$ $\frac{5}{8}$ $\frac{7}{9}$

Copyright by Goodheart-Willcox Co., Inc.

Name

Lettering Practice

Date

Problem

Date

Name

Complete the missing view.

Complete the missing view.

Name		Date	Problem

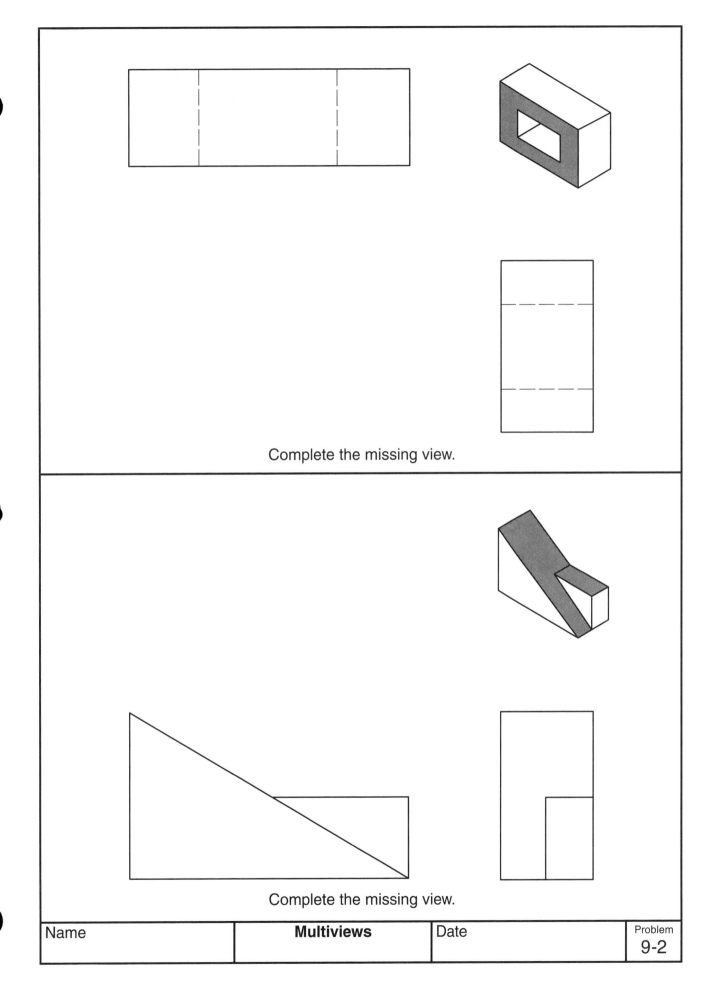

Complete the missing view.

Complete the missing view.

| Name | **Multiviews** | Date | Problem 9-2 |

Name		Date	Problem

Complete the missing view.

Complete the missing view.

Name		Date	Problem

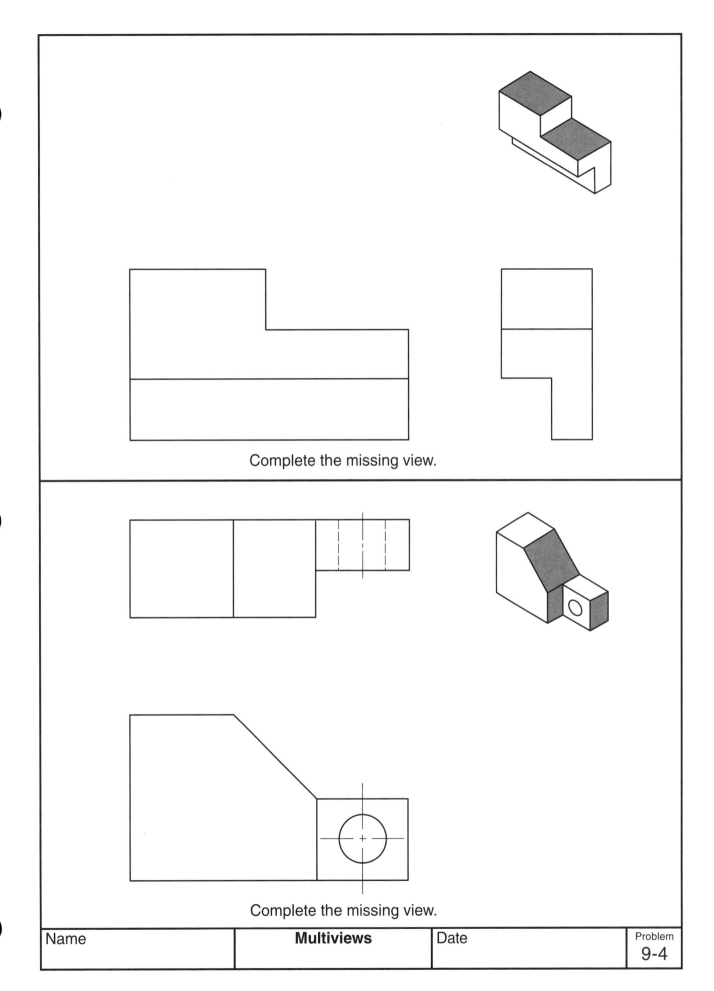

Complete the missing view.

Complete the missing view.

Name	**Multiviews**	Date	Problem 9-4

Name		Date	Problem

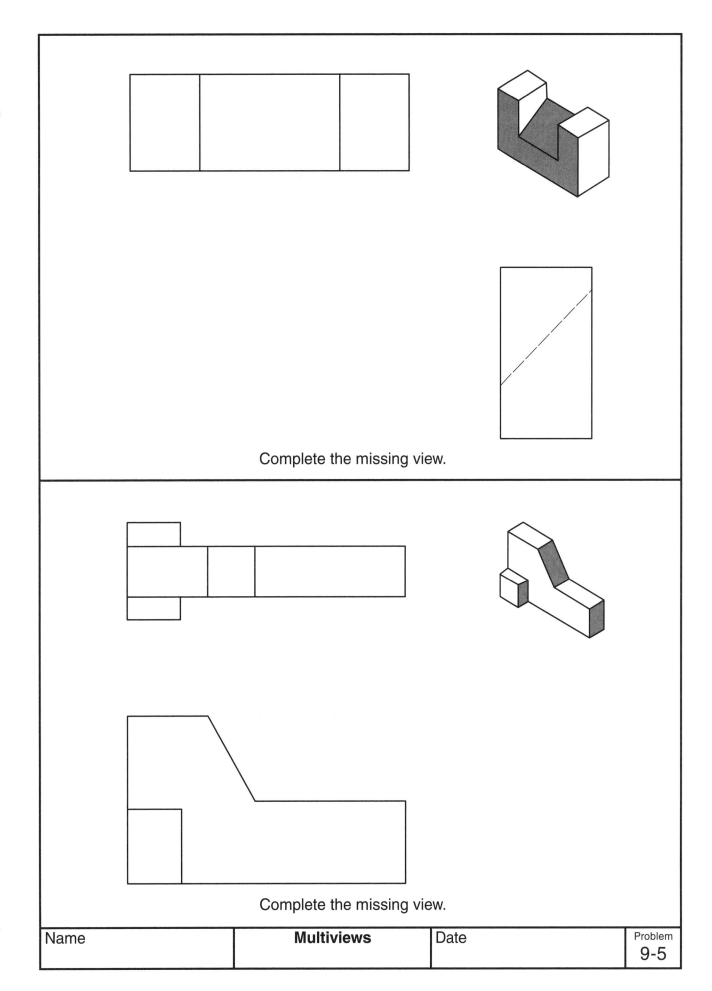

Complete the missing view.

Complete the missing view.

| Name | **Multiviews** | Date | Problem 9-5 |

Name		Date	Problem

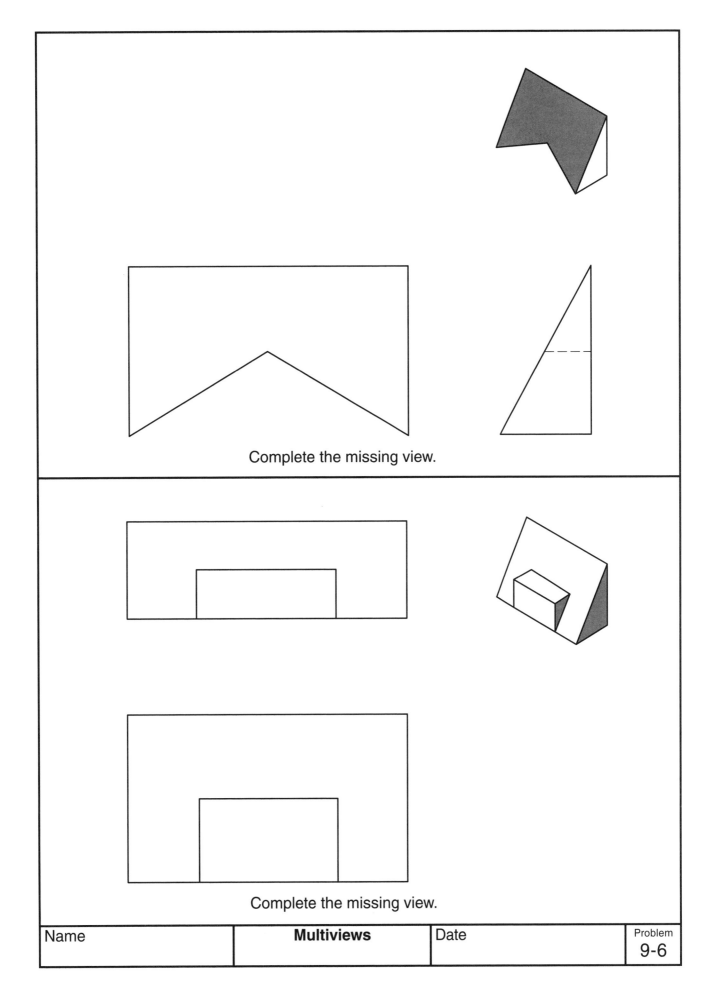

Complete the missing view.

Complete the missing view.

| Name | **Multiviews** | Date | Problem
9-6 |

Name		Date	Problem

Draw the views necessary to describe the object.

| Name | Yoke | Date | Problem 9-7 |

Problem

Date

Name

25.0

35.0

50.0

40.0

100.0

60.0

25.0

Dimensions are in MM

Draw the views necessary to describe the object.

Name

Step Block

Date

Problem
9-8

Problem

Date

Name

5

$1\frac{1}{2}$

1

$\frac{1}{2}$

$\frac{1}{2}$

$2\frac{1}{2}$

Date

Slide

Draw the views necessary to describe the object.

Name

Problem

Date

Name

Date

Guide

Draw the views necessary to describe the object.

Name

Problem

Date

Name

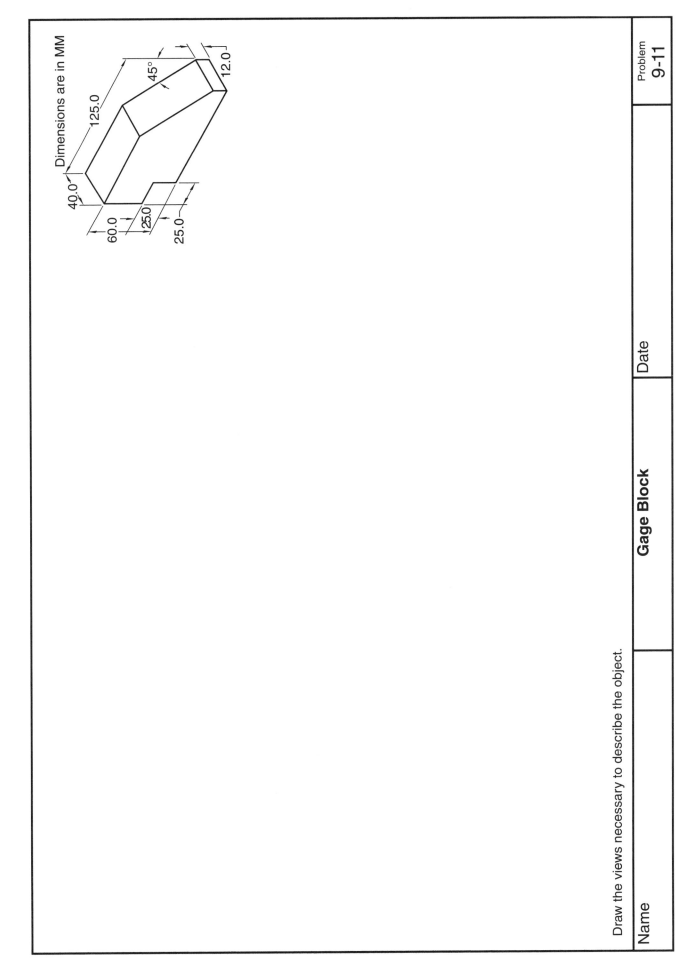

Dimensions are in MM

125.0

45°

12.0

40.0

60.0

25.0

25.0

Draw the views necessary to describe the object.

Gage Block

Name

Date

Problem

Date

Name

Date

Aligner

Draw the views necessary to describe the object.

Name

Problem

Date

Name

Ø2

Ø.75 THRU

4

Draw the views necessary to describe the object.

Name

Bearing

Date

Name

Date

Problem

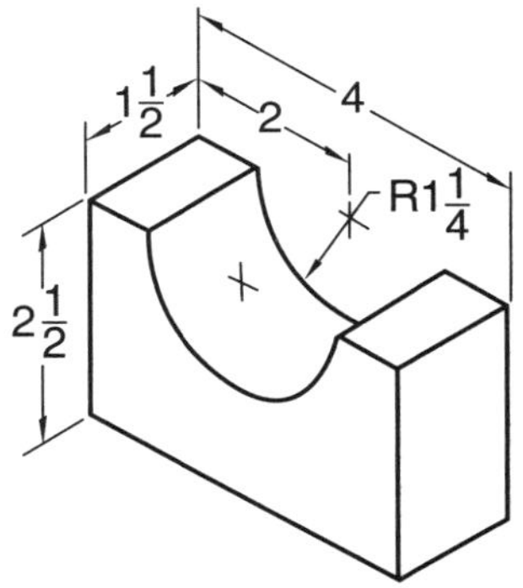

Draw the views necessary to describe the object.

Name	**Guide Block**	Date	Problem 9-14

Problem

Date

Name

Draw the views necessary to describe the object.

Name

Balance

Date

Problem

Date

Name

45°

$1\frac{1}{2}$

4

$\frac{3}{4}$

$2\frac{1}{2}$

$\frac{3}{4}$

$\frac{1}{2}$ (TYP.)

Draw the views necessary to describe the object.

Name

Angled Guide

Date

Problem
9-16

Problem

Date

Name

Draw the views necessary to describe the object.

Template Block

Name

Date

Problem

Date

Name

2X ⌀.875
1.5 DEEP
R1.25
1.5

Date

Link

Draw the views necessary to describe the object.

Name

Problem

Date

Name

Draw the views necessary to describe the object.

Name

Bracket

Date

Problem

Date

Name

4X ⌀.312
EQ SP

⌀4.75
.75 ⌀4.25
1

⌀3.5

⌀1.25

⌀2.0

1

Draw the views necessary to describe the object.

Name

Flange

Date

Problem

Date

Name

Complete the missing view.

Advanced Problem

Name

Date

Problem

Date

Name

Complete the missing view.

Date

Advanced Problem

Name

Problem

Date

Name

Dimension each problem correctly.

Name

Date

Dimensioning Problems

Problem
10-1

Problem

Date

Name

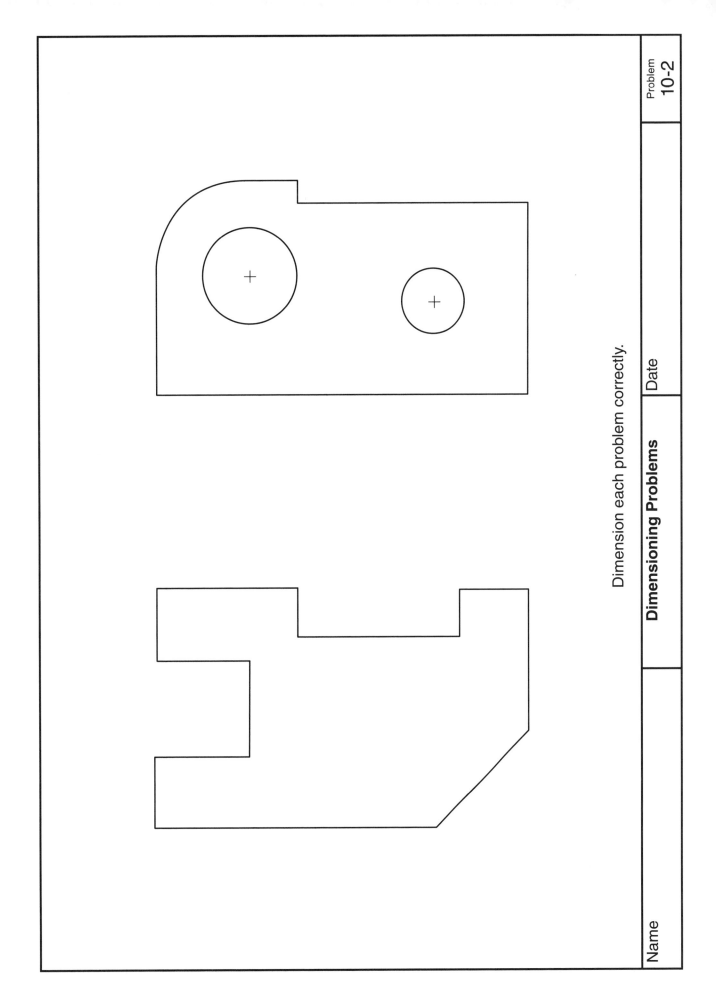

Dimension each problem correctly.

Date

Dimensioning Problems

Name

Problem

Date

Name

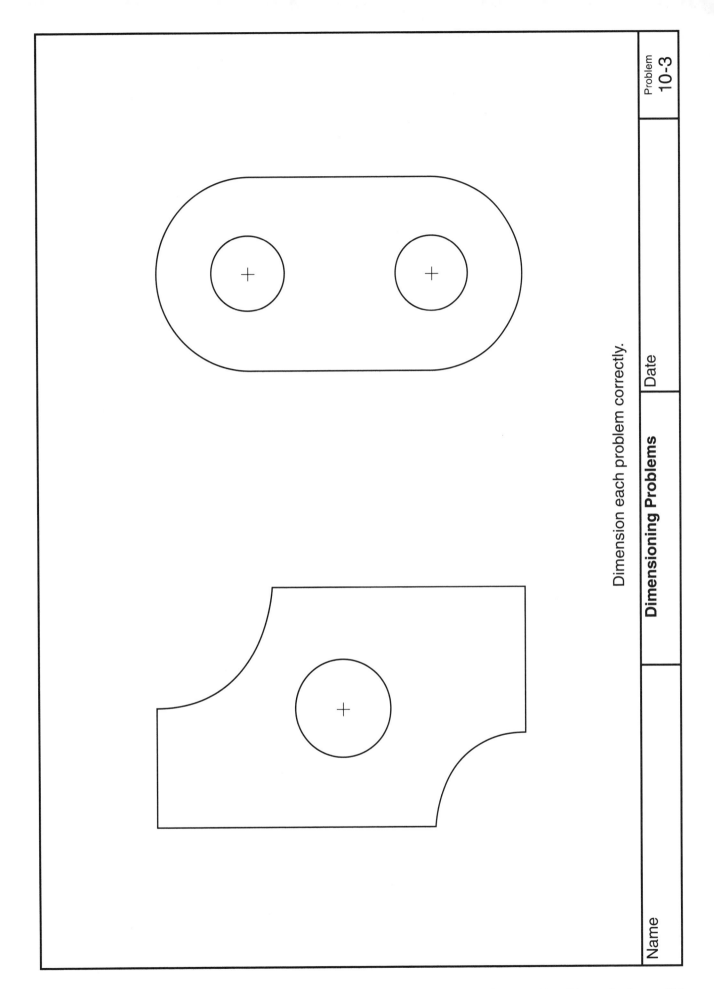

Dimension each problem correctly.

Date

Dimensioning Problems

Name

Problem

Date

Name

Dimension each problem correctly.

Dimensioning Problems

Name

Date

Problem

Date

Name

Dimension each problem correctly.

Date

Dimensioning Problems

Name

Problem

Date

Name

Dimension the view correctly.

Date

Dimensioning Problems

Name

Problem

Date

Name

Aluminum and Magnesium

Rubber and Plastic

Wood (End Grain)

Steel

White Metal, Zinc, and Lead

Wood (with Grain)

Cast Iron and General Purpose

Brass, Bronze, and Copper

Concrete

Draw the standard section lines for the material listed.

Sectioning Symbols

Name		Date	Problem 11-1

Problem

Date

Name

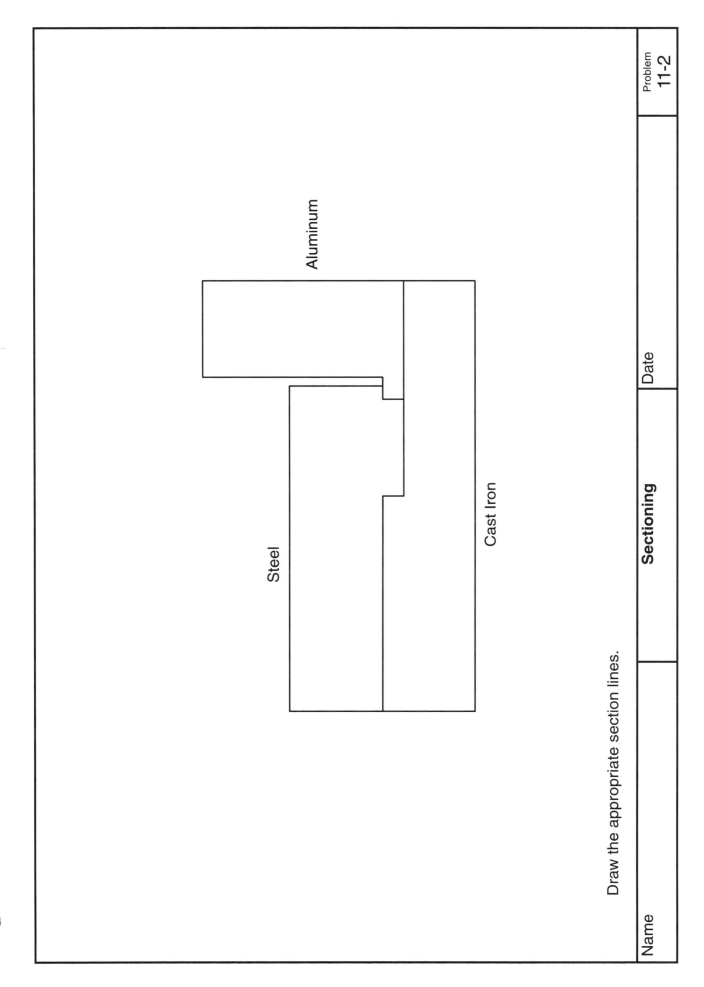

Draw the appropriate section lines.

Aluminum

Steel

Cast Iron

Name

Sectioning

Date

Problem
11-2

Problem

Date

Name

Complete the views necessary to describe a full section of the object. Draw a cutting-plane line and section lines as needed. Refer to the dimensions given in the text and dimension the drawing.

Grinding Wheel

Name

Date

Problem
11-3

Problem

Date

Name

Complete the views necessary to describe a full section of the object. Draw a cutting-plane line and section lines as needed. Refer to the dimensions given in the text and dimension the drawing.

Grinding Wheel

Name _____   Date _____

Problem

Date

Name

Complete the views necessary to describe a half section of the object. Refer to Problem P11-5 in the text. Draw a cutting-plane line, object lines, and section lines as needed. Refer to the dimensions given in the text and dimension the drawing.

Grinding Wheel

Name

Date

Problem
11-5

Problem

Date

Name

Draw a front view and full section of the object. Refer to Problem P11-7 in the text. Draw a cutting-plane line, object lines, and section lines as needed. Refer to the dimensions given in the text and dimension the drawing.

Name		Coupling	Date	Problem 11-6

Problem

Date

Name

Draw a front view and half section of the object. Refer to Problem P11-10 in the text. Draw a cutting-plane line, object lines, and section lines as needed. Refer to the dimensions given in the text.

Name		Date	Problem
	Flat Belt Pulley		**11-7**

Problem

Date

Name

Draw the views needed to show the shape of the spacer. Draw the right-side view as an offset section through the three holes. Draw the cutting-plane line on the primary view and dimension the drawing.

Name

Spacer

Date

Problem 11-8

Problem

Date

Name

2X Ø1.50

2X Ø.75

4.50

.125

.25

.187

.187

1.00

.75

.25

1.25

Draw the views needed to show the shape of the rod. Show the rod's cross section as a revolved section. Draw the cutting-plane line on the primary view and dimension the drawing.

Name		Date	Problem
	Connecting Rod		**11-9**

Problem

Date

Name

.125 × .250 KEYWAY
Ø1.00 REAM
5× Ø.50 DRILL
EQ SP
1.00
Ø3.00
Ø4.00

Draw the views needed to show the shape of the plate. Draw the right-side view as an aligned section and dimension the drawing.

Adapter Plate

Name

Date

Problem

Date

Name

Draw an auxiliary view. Do not dimension.

Mitered Extrusion

Date

Name

Problem

Date

Name

Draw an auxiliary view. Do not dimension.

Name

Notched Block

Date

Problem

Date

Name

Draw an auxiliary view. Do not dimension.

Chapter 12   Auxiliary Views   **143**

Problem
12-3

Date

Truncated Hexagon

Name

Problem

Date

Name

Draw an auxiliary view. Do not dimension.

Problem

12-4

Date

Truncated Octagon

Name

Problem

Date

Name

Draw an auxiliary view. Do not dimension.

Angled Wedge

Date

Name

Problem

Date

Name

Draw an auxiliary view. Do not dimension.

Problem
12-6

Date

Truncated Cylinder

Name

Problem

Date

Name

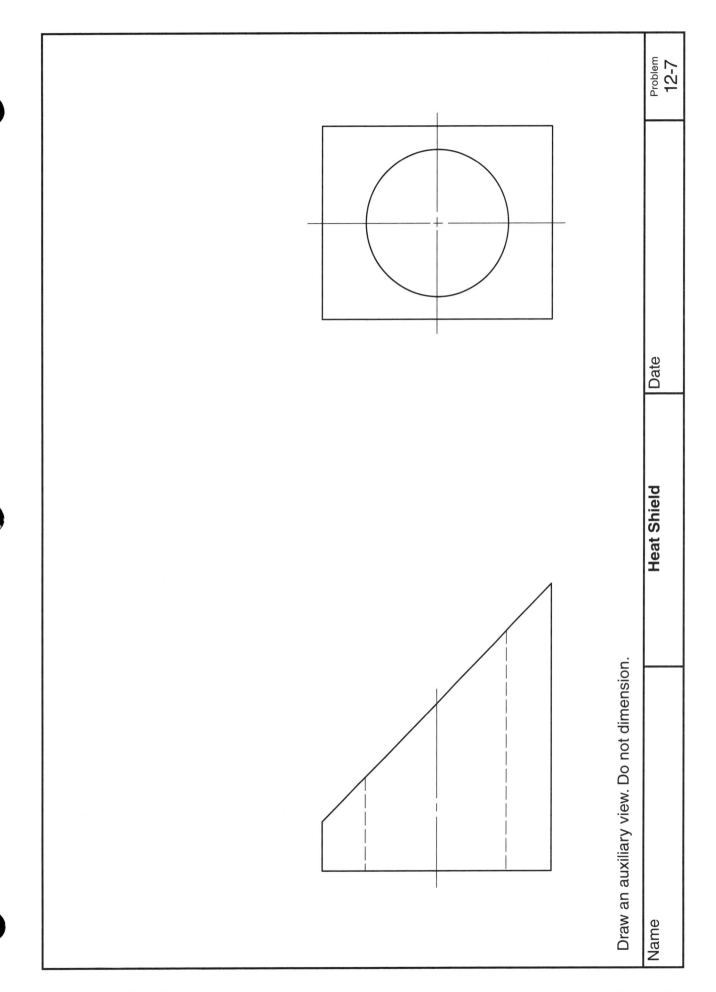

Draw an auxiliary view. Do not dimension.

Date

Heat Shield

Name

Problem

Date

Name

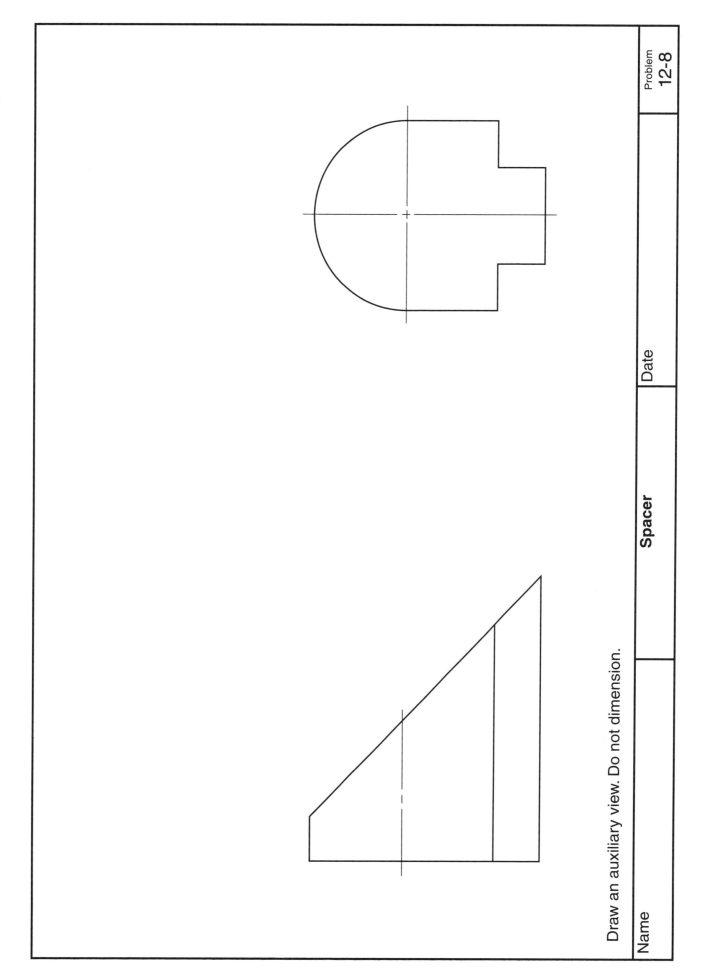

Draw an auxiliary view. Do not dimension.

Name

Spacer

Date

Problem

Date

Name

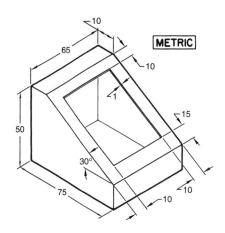

METRIC

Draw the orthographic and auxiliary views needed to describe the object. Do not dimension.

Name	**Instrument Case**	Date	Problem 12-9

Name		Date	Problem

Draw the orthographic and auxiliary views needed to describe the object. Do not dimension.

$R\frac{3}{4}$

$2\times\varnothing\frac{1}{2}$

$\frac{15}{16}$

$\frac{15}{16}$

$\frac{15}{16}$

$2\frac{7}{16}$

$30°$

$\frac{3}{8}\times45°$

$\frac{3}{4}$

$\frac{3}{8}$

$\frac{1}{2}$

$1\frac{1}{8}$

$\frac{3}{16}$

Name

Bracket

Date

Problem

Date

Name

Draw the orthographic and auxiliary views needed to describe the object. Do not dimension.

2X Ø.75

1.5

1.5

.75

.75

1.0

1.0

.5

.5

.75

.5

.75

1.5

37°

37°

45°

.75

.187

.75

.75

Name

Date

Hanger Clamp

Problem

Date

Name

Draw the orthographic and auxiliary views needed to describe the object. Do not dimension.

Ø.437 THRU

Ø1.00

1.50

.50

1.00

.25

.50

.50

45°

.62

3.00

Name

Date

Shifter Bar

Problem
12-12

Problem

Date

Name

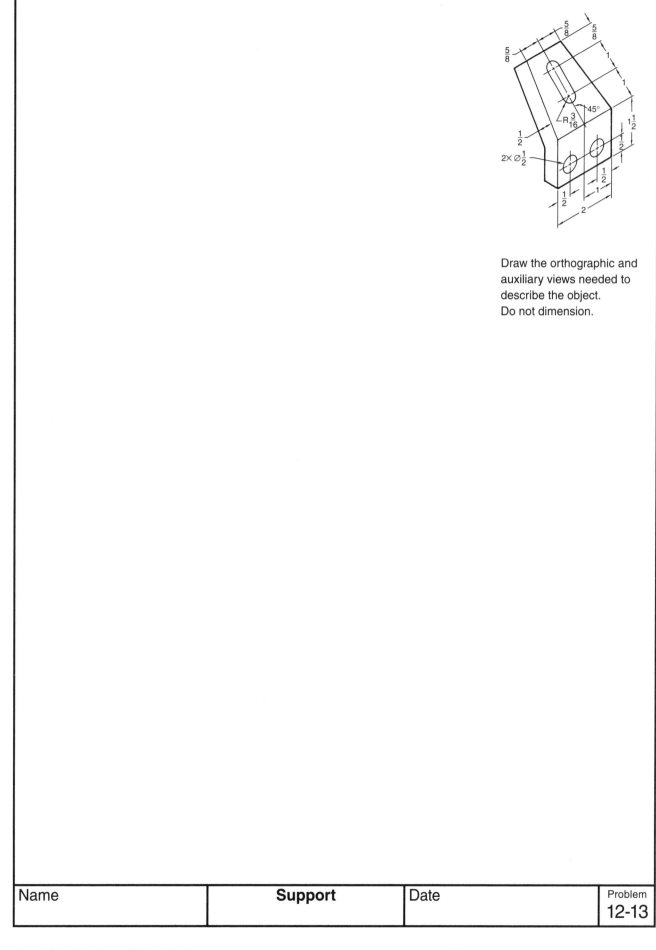

Draw the orthographic and
auxiliary views needed to
describe the object.
Do not dimension.

| Name | **Support** | Date | Problem
12-13 |

Name		Date	Problem

Draw the orthographic and auxiliary views needed to describe the object. Do not dimension.

R.25 (TYP.)

2.5

1.0

.62

62

.75

.312 × 45°

2.0

.75

.50

135°

.50

.50

2.0

1.0

4× Ø.375

Name

Date

Adjustable Bracket

Problem
12-14

Copyright by Goodheart-Willcox Co., Inc.

Chapter 12   Auxiliary Views   165

Problem

Date

Name

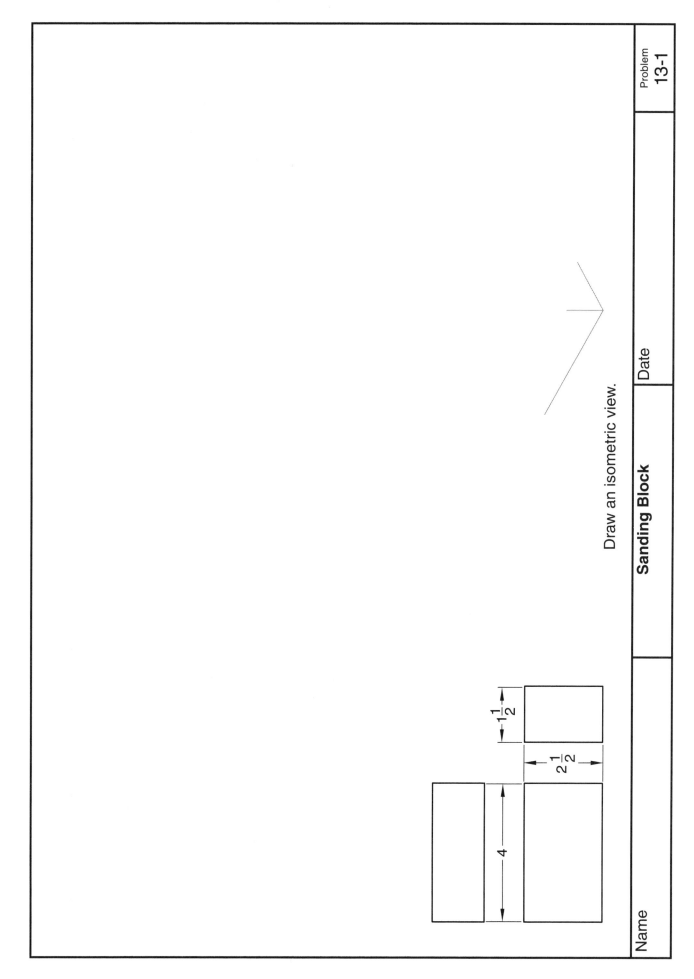

Draw an isometric view.

Name

Sanding Block

Date

Problem

Date

Name

Draw an isometric view and dimension.

Doorstop

2

$1\frac{1}{2}$

5

Date

Name

Problem

Date

Name

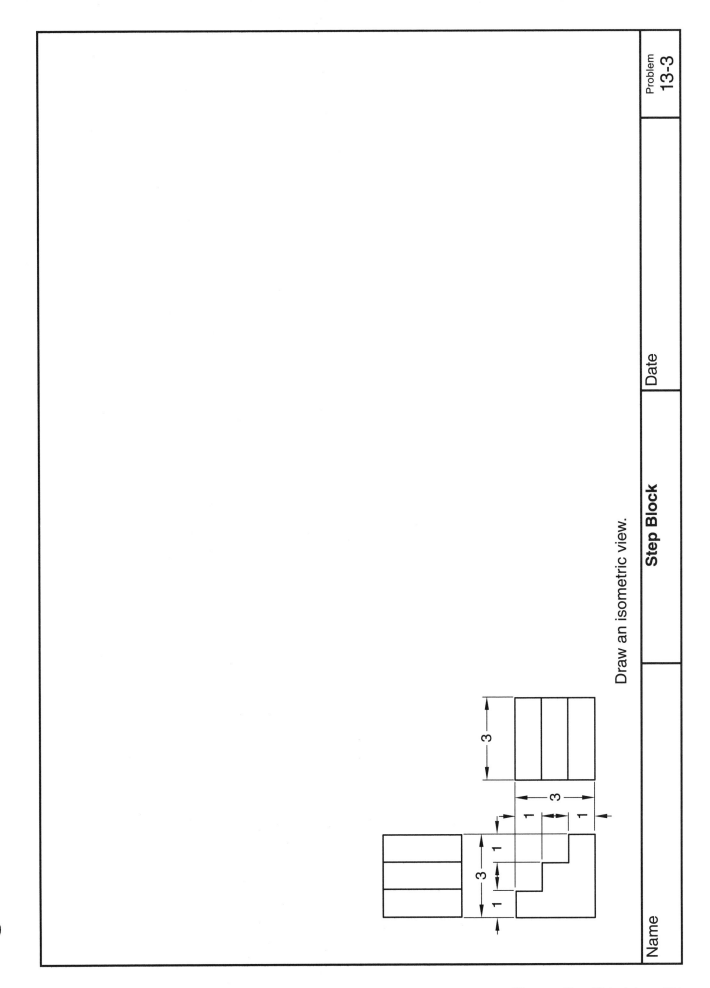

Draw an isometric view.

Name	Step Block	Date	Problem 13-3

Problem

Date

Name

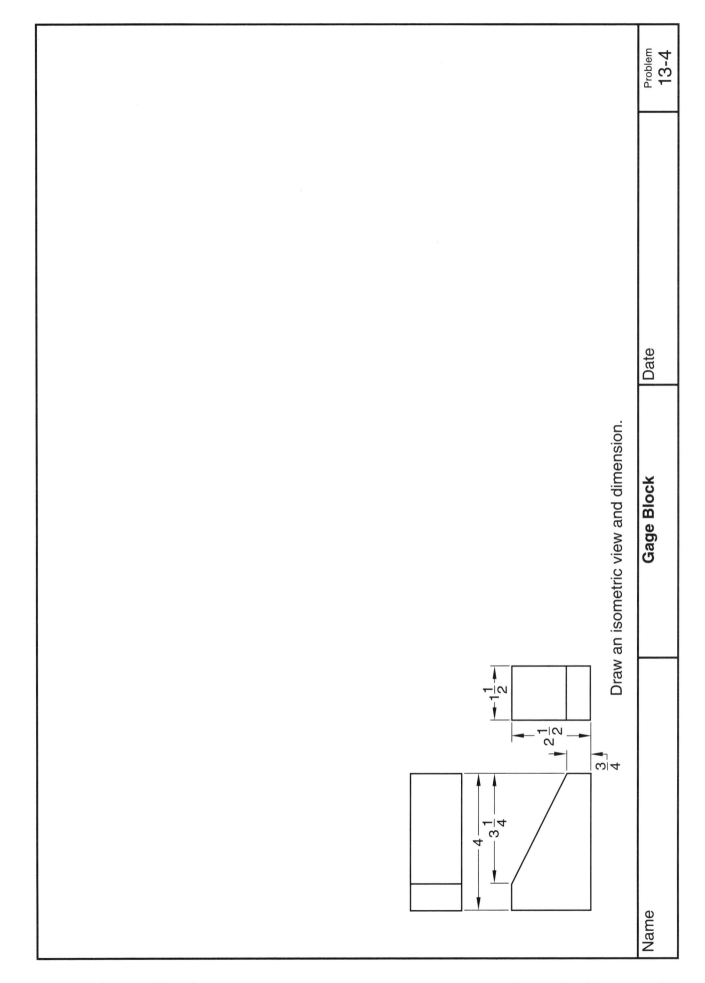

Draw an isometric view and dimension.

| Name | | Gage Block | Date | Problem 13-4 |

Problem

Date

Name

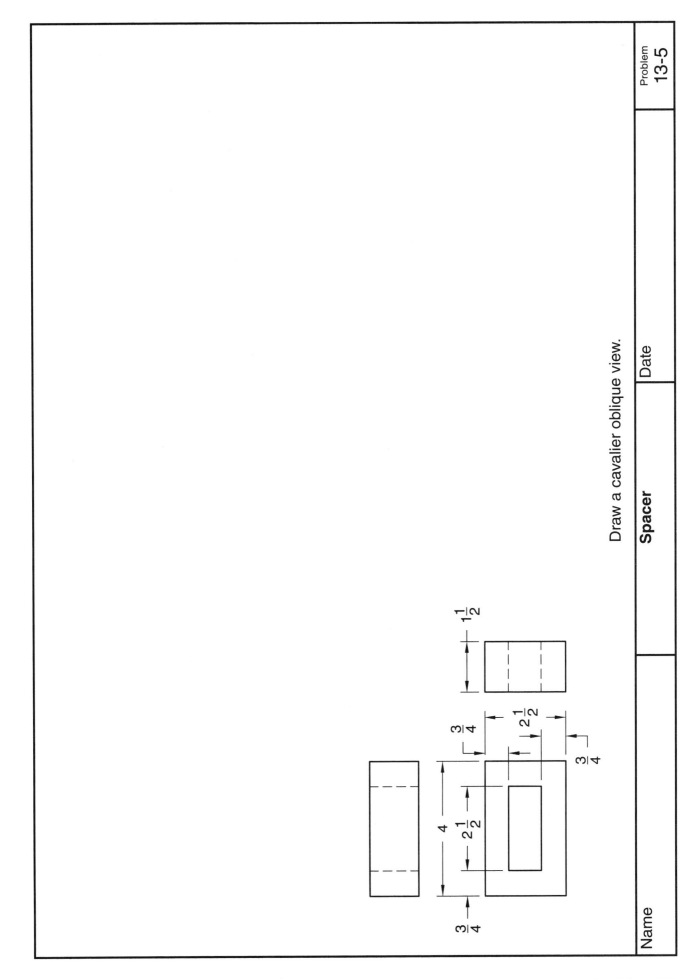

Draw a cavalier oblique view.

Spacer

Name

Date

Problem

Date

Name

Draw an isometric view.

$1\frac{1}{2}$

$\frac{3}{4}$

$2\frac{1}{2}$

4

$1\frac{3}{4}$

$1\frac{1}{8}$

Name

Slide

Date

Problem

13-6

Problem

Date

Name

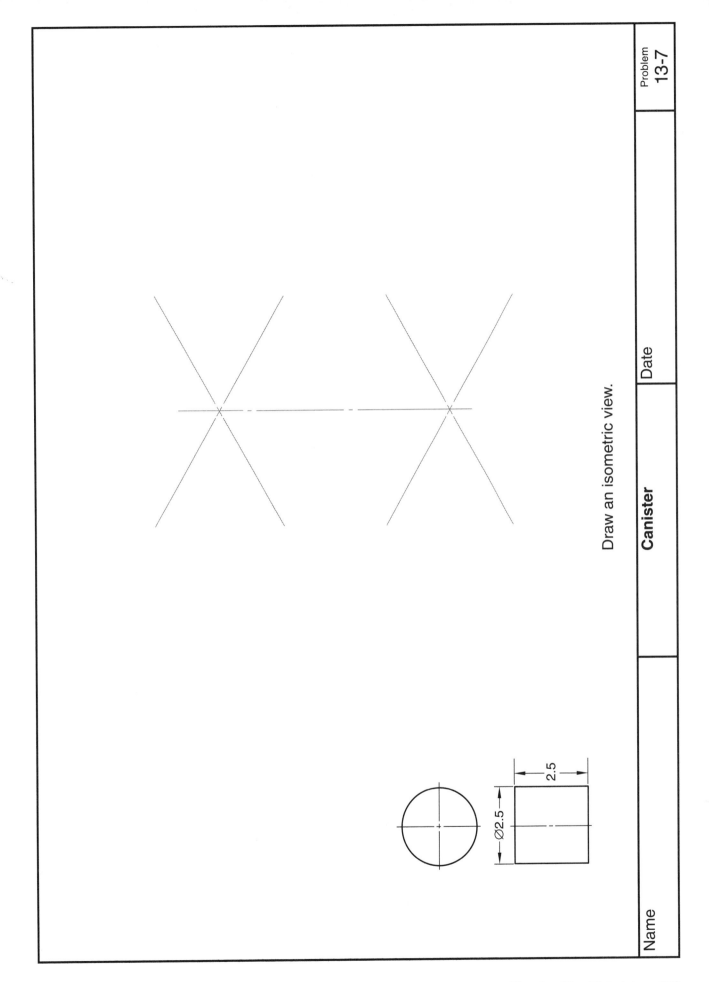

Draw an isometric view.

Ø2.5

2.5

Date

Canister

Name

Problem

Date

Name

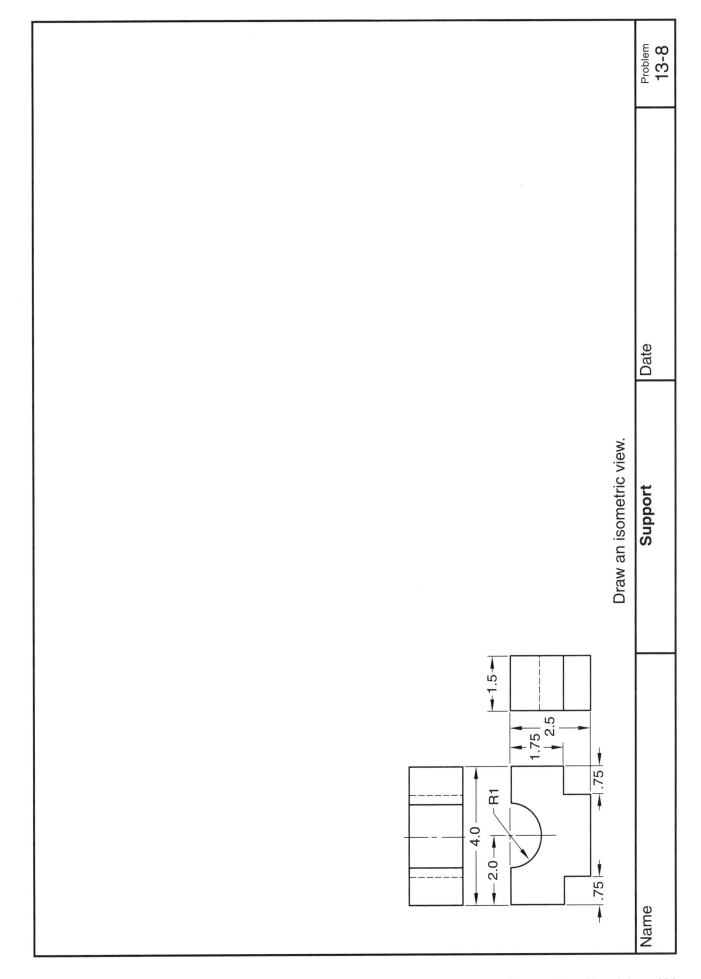

Draw an isometric view.

Date

Support

Name

Problem

Date

Name

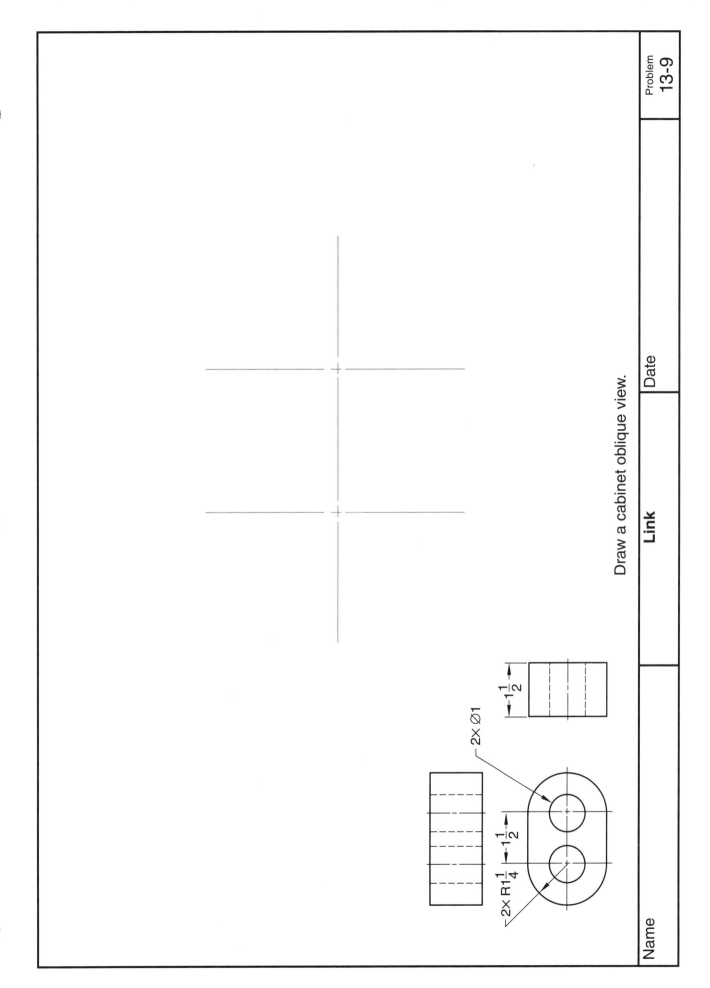

Draw a cabinet oblique view.

2× Ø1

2× R1¼

1½

1½

Problem 13-9

Name

Link

Date

Problem

Date

Name

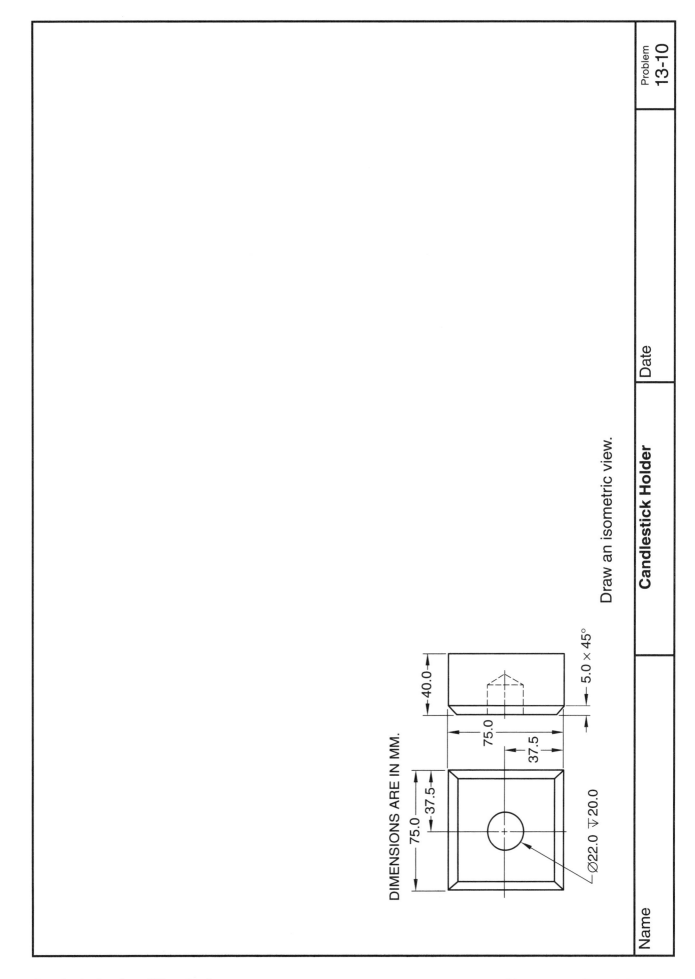

DIMENSIONS ARE IN MM.

Ø22.0 ⫤20.0

75.0

37.5

75.0

37.5

40.0

5.0 × 45°

Draw an isometric view.

Name | Date | **Candlestick Holder** | Problem **13-10**

Problem

Date

Name

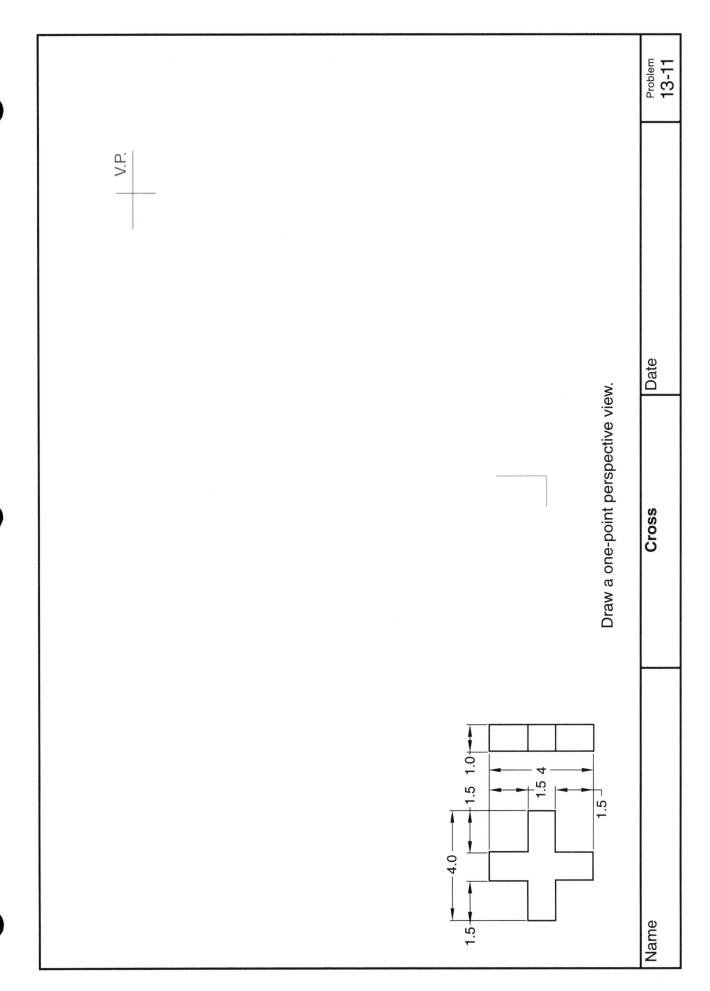

V.P.

Draw a one-point perspective view.

| Name | Cross | Date | Problem 13-11 |

Problem

Date

Name

Draw a pattern for the cylinder.

Name

Cylinder

Date

Problem

Date

Name

Draw a pattern for the prism.
Allow material for the seams.

Problem
14-2

Date

Rectangular Prism

Name

Problem

Date

Name

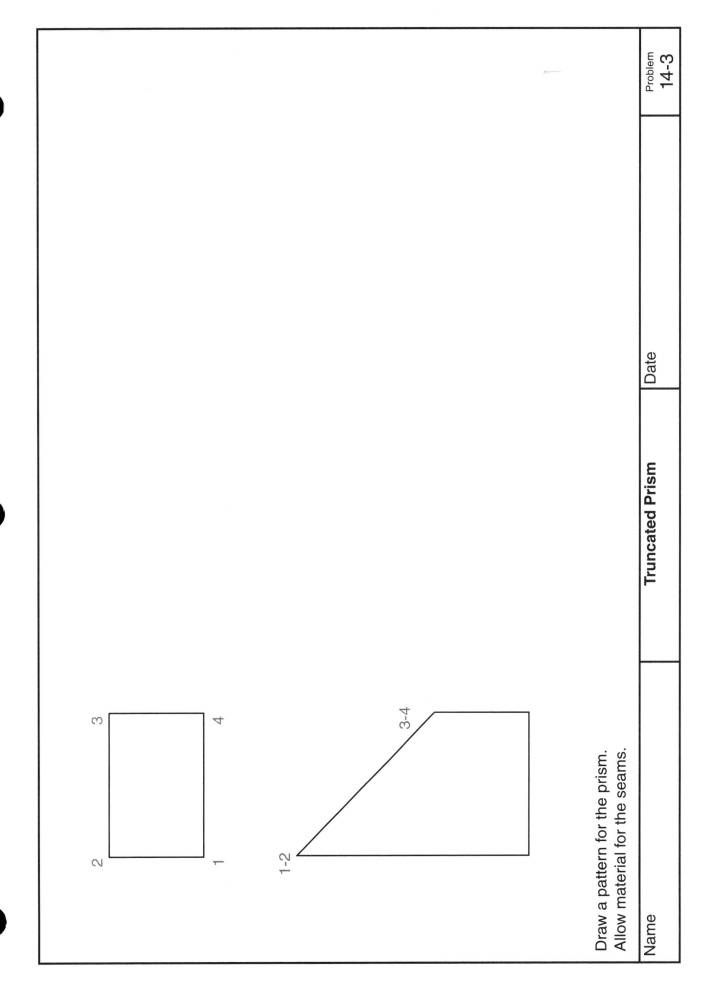

Draw a pattern for the prism.
Allow material for the seams.

Truncated Prism

Name

Date

Problem

Date

Name

Draw a pattern for the truncated cylinder.

Name

Date

Truncated Cylinder

Problem

Date

Name

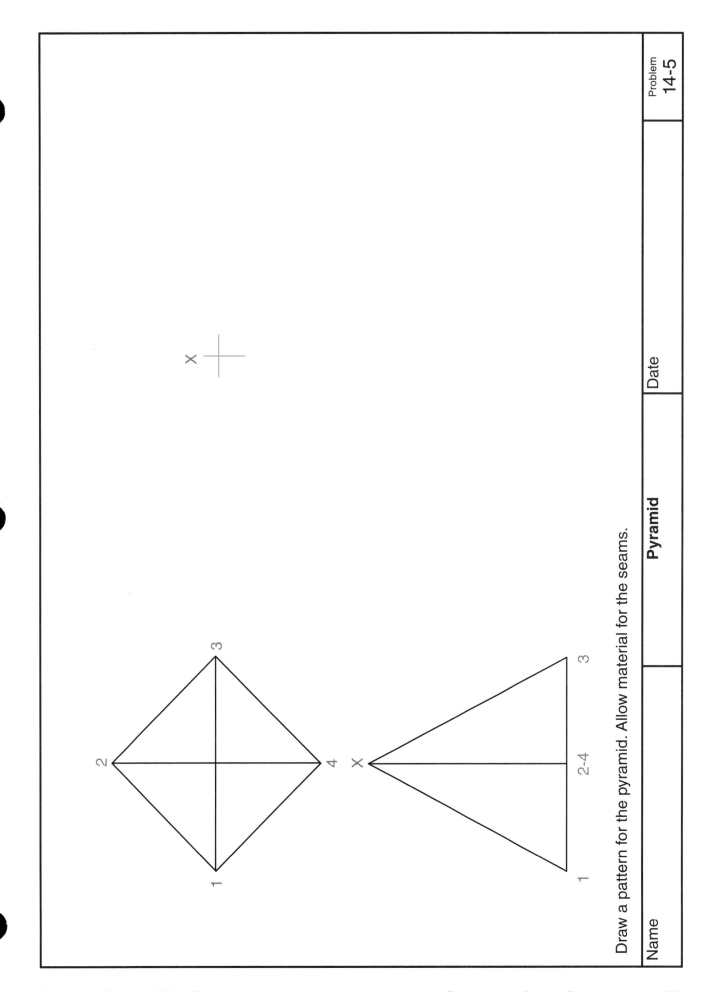

Draw a pattern for the pyramid. Allow material for the seams.

Name		**Pyramid**	Date
			Problem 14-5

Problem

Date

Name

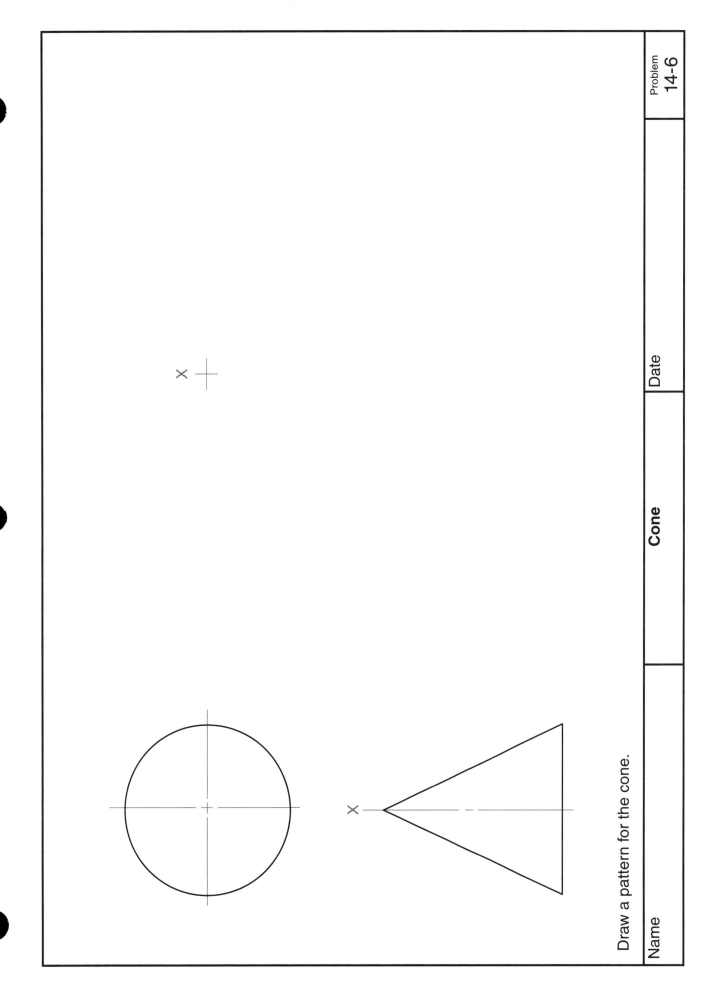

Draw a pattern for the cone.

Problem
14-6

Date

Cone

Name

Problem

Date

Name

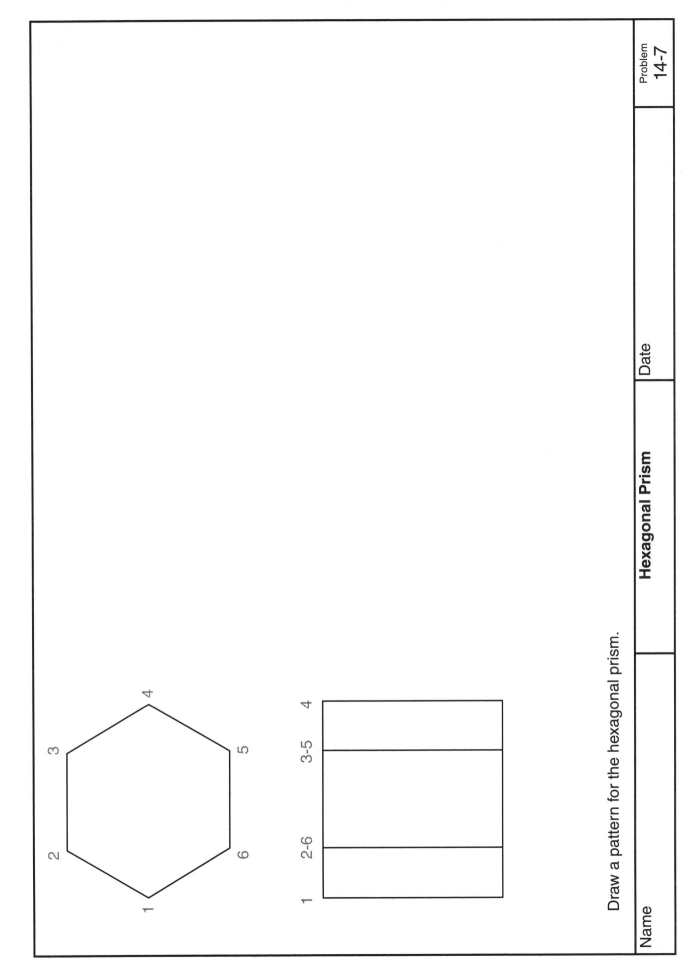

Draw a pattern for the hexagonal prism.

Hexagonal Prism

Name

Date

Problem

Date

Name

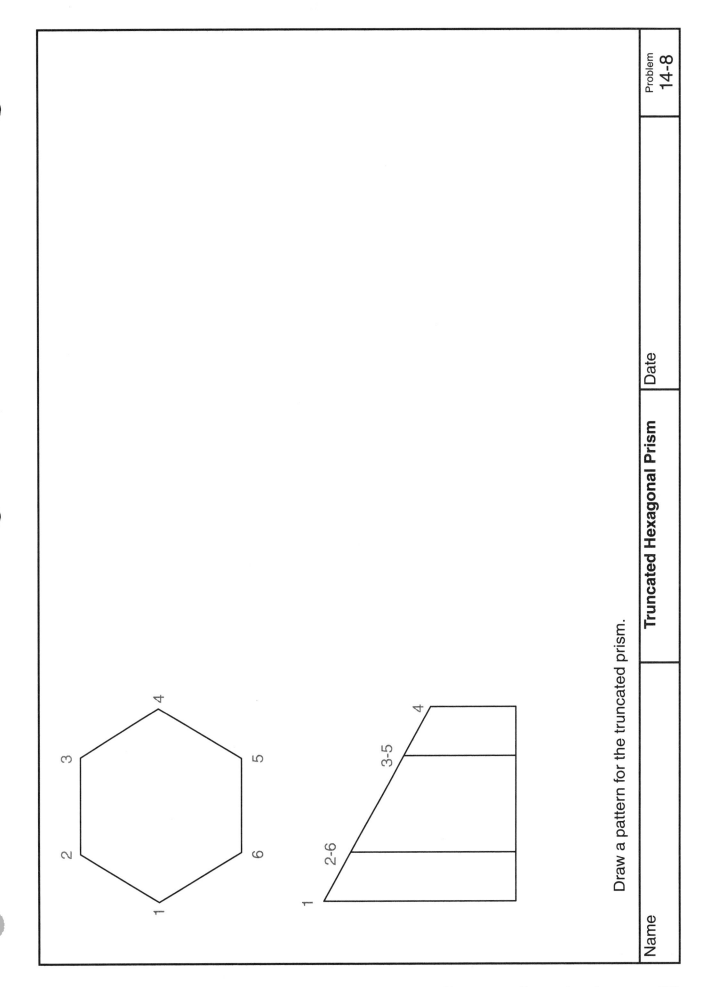

Draw a pattern for the truncated prism.

Name _____

Truncated Hexagonal Prism

Date _____

Problem

Date

Name

Refer to Problem 15-1 in the text and make a detail drawing of the machinist's square.

Machinist's Square Details		Date

Name

Problem

Date

Name

Refer to Problem 15-5 in the text and make an assembly drawing of the C-clamp.

Name	C-Clamp Assembly		Date	Problem 15-2

Problem

Date

Name

Refer to Problem 15-7 in the text and prepare a parts list and bill of materials for the deck gun.

Name	Deck Gun Information	Date	Problem 15-3

Problem

Date

Name

Design a sports car using the layout parameters shown. Add racing stripes. Use colored pencils.

Name	Design Problem	Date	Problem 17-1

Problem

Date

Name

Design a hand-launched glider in the space above. Construct a balsa model to verify flying characteristics.

Name		Design Problem	Date	Problem 17-2

Problem

Date

Name

Draw a map of the school grounds at your school.

School Grounds

Name			Date	Problem 19-1

Problem

Date

Name

Draw a plot plan of the property on which your home is located.

| | **Plot Plan** | Date |

Name

Problem

Date

Name

Draw a map of your neighborhood.

Name		Neighborhood Map	Date

Problem

Date

Name

Name

Date

Pie Graph

Create a pie graph that shows how you spend your earnings or allowance.

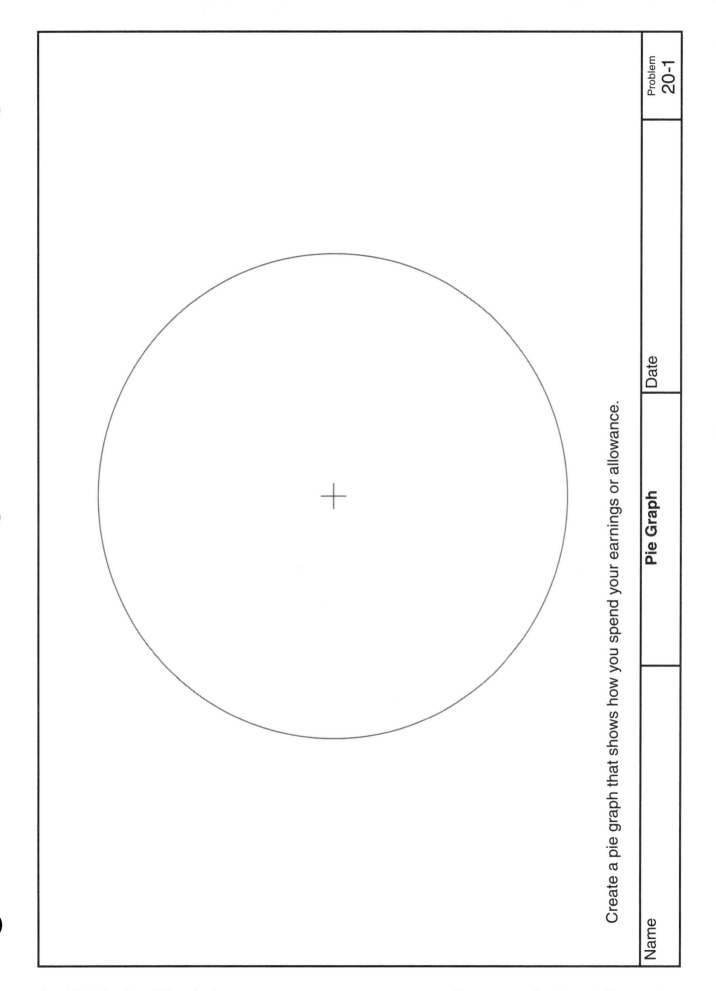

Problem

Date

Name

Create a bar graph showing the change in automobile engine horsepower from 1940 to the present. Use five-year steps.

Bar Graph

Name		Date	Problem 20-2

Problem

Date

Name

Create a line graph showing how the price of the basic automobile has increased since 1940. Use five-year steps.

Line Graph

Name		Date		Problem
				20-3

Problem

Date

Name

Create a pictorial graph showing the enrollment in each grade of your school. Let each symbol represent 25 students.

Picture Graph		
Name	Date	Problem 20-4

Problem

Date

Name

3X Ø$\frac{1}{2}$

3X R$\frac{3}{4}$

2$\frac{3}{8}$

1

$\frac{3}{16}$ FILLET WELDS

BRACKET

$\frac{3}{8}$ STEEL PLATE

Draw the orthographic views necessary to describe the assembly and fabrication of the object. Use the correct symbols for the welding information specified. Welds are to be made on both sides of the joint.

Name		Date	
	Bracket		Problem **21-1**

Problem

Date

Name

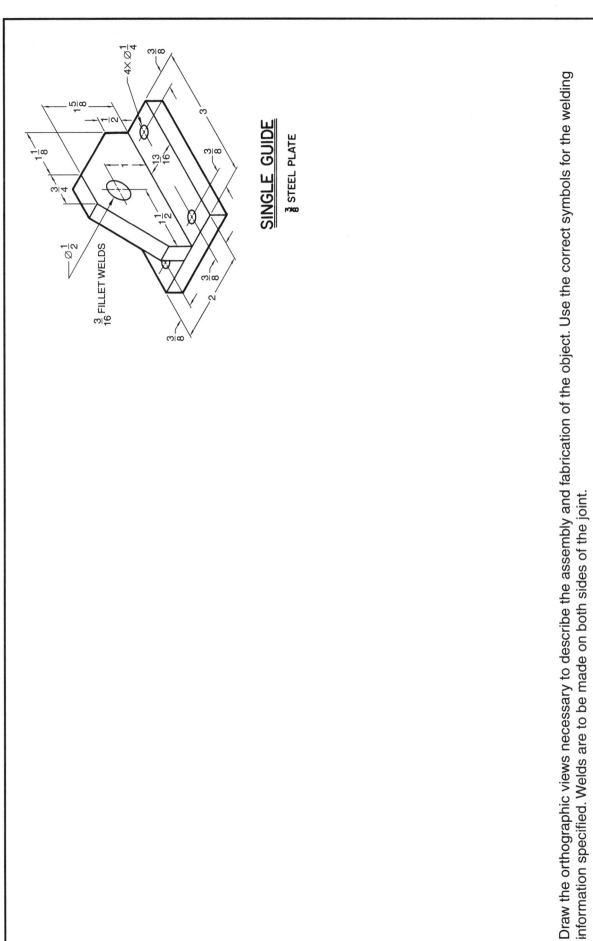

SINGLE GUIDE
$\frac{3}{8}$ STEEL PLATE

Draw the orthographic views necessary to describe the assembly and fabrication of the object. Use the correct symbols for the welding information specified. Welds are to be made on both sides of the joint.

Name		Single Guide	Date	Problem 21-2

Problem

Date

Name

PIVOT
C.F. STEEL

∅1½

∅1⅛

4× ∅¼

3/16 FILLET WELD

1

1

1

3/8

3/8

3/8

1¾

2½

3/8

1¾

2½

Draw the orthographic views necessary to describe the assembly and fabrication of the object. Use the correct symbols for the welding information specified. The weld is to be made all around the joint.

Name		Pivot	Date

Problem

Date

Name

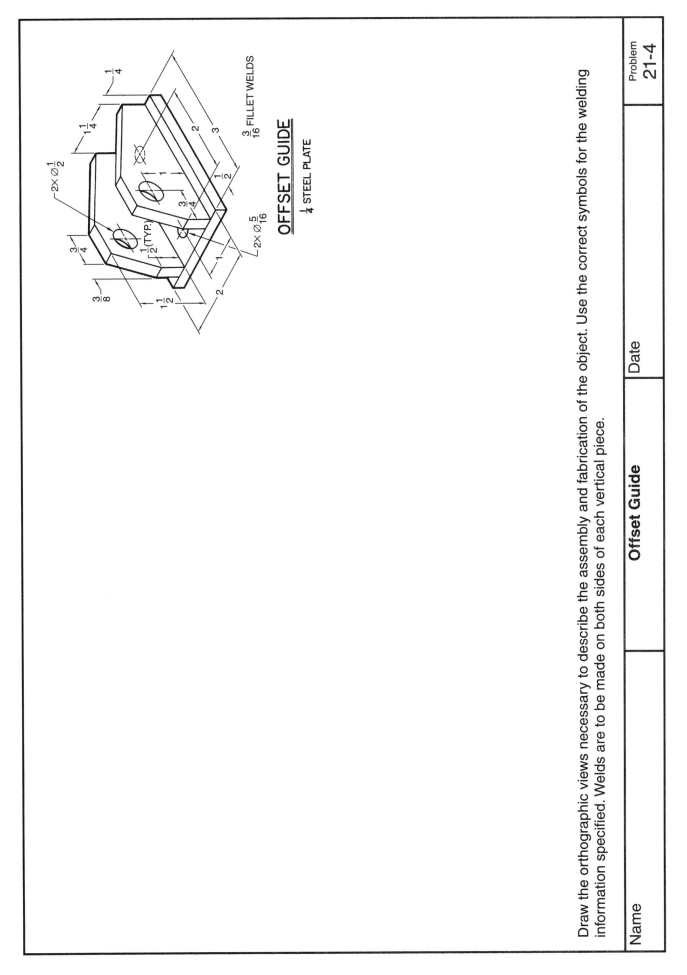

OFFSET GUIDE

$\frac{1}{4}$ STEEL PLATE

$\frac{3}{16}$ FILLET WELDS

$2\times \emptyset \frac{1}{2}$

$2\times \emptyset \frac{5}{16}$

$1\frac{1}{4}$

$\frac{1}{4}$

$\frac{3}{8}$

$1\frac{1}{2}$

$\frac{3}{4}$

$\frac{1}{2}$(TYP.)

2

3

2

1

$\frac{3}{4}$

1

$\frac{1}{2}$

3

2

1

Draw the orthographic views necessary to describe the assembly and fabrication of the object. Use the correct symbols for the welding information specified. Welds are to be made on both sides of each vertical piece.

Name		**Offset Guide**	Date	Problem 21-4

Problem

Date

Name

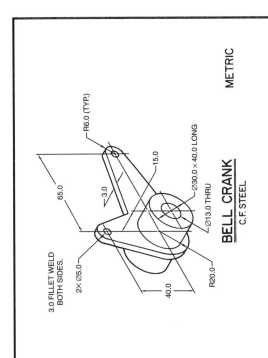

METRIC

R6.0 (TYP.)

65.0

3.0

15.0

∅30.0 × 40.0 LONG

∅13.0 THRU

BELL CRANK
C.F. STEEL

3.0 FILLET WELD
BOTH SIDES.

2× ∅5.0

40.0

R20.0

Draw the orthographic views necessary to describe the assembly and fabrication of the object. Use the correct symbols for the welding information specified. A 3 mm weld is to be made on both sides of the joint.

Name		**Bell Crank**	Date	Problem **21-5**

Problem

Date

Name

Draw a 4" long hexagonal head bolt and nut with 3/4-10UNC-2 threads. Draw a 4" long square head bolt and nut with the same thread specifications. Use simplified thread representations.

Name	Date	Problem
		22-1
Bolts and Nuts		

Problem

Date

Name

Draw a 3" long hexagonal head bolt and nut with 1-8UNC-2 threads. Draw a 3" long square head bolt and nut with the same thread specifications. Use schematic thread representations.

Name		Date	Problem 22-2
	Bolts and Nuts		

Problem

Date

Name

Symbols

Plug

Lightbulb

Switch

Make a schematic diagram of a desk lamp.

| Name | | **Desk Lamp Schematic** | Date | Problem 23-1 |

Problem

Date

Name

Prepare a schematic diagram of a two-cell flashlight.

Flashlight Schematic

Name

Date

Problem
23-2

Problem

Date

Name

Draw a wiring diagram of your bedroom.

Name		Wiring Diagram	Date	Problem 23-3

Problem

Date

Name

Make a scale drawing of the school drafting room. Use a scale of 1/4" = 1'-0".

Name	Room Layout	Date

Problem
24-1

Problem

Date

Name

Draw a floor plan of your home or apartment.

Name		Floor Plan	Date	Problem 24-2

Copyright by Goodheart-Willcox Co., Inc.

Chapter 24 Architectural Drafting **251**

Problem

Date

Name

Design and draw a floor plan for a small vacation cabin.

Vacation Cabin

Name

Date

Problem

Date

Name

Problem

Date

Name

Problem

Date

Name